Getting Around School

Sandy Phan

Consultants

Shelley Scudder
Gifted Education Teacher
Broward County Schools

Caryn Williams, M.S.Ed.
Madison County Schools
Huntsville, AL

Publishing Credits

Dona Herweck Rice, *Editor-in-Chief*
Lee Aucoin, *Creative Director*
Torrey Maloof, *Editor*
Diana Kenney, M.A.Ed., NBCT,
Associate Education Editor
Marissa Rodriguez, *Designer*
Stephanie Reid, *Photo Editor*
Rachelle Cracchiolo, M.S.Ed., Publisher

Image Credits: Cover & pp. 1, 11, 13, 19, 21 Alamy; p. 6 BigStock; pp. 3, 5, 18 iStockphoto; p. 14 Lee Aucoin; p. 12 The Library of Congress [LC-USE6-D-006302]; pp. 8–10, 16, 24 Mapping Specialists; All other images from Shutterstock.

Teacher Created Materials
5301 Oceanus Drive
Huntington Beach, CA 92649-1030
http://www.tcmpub.com

ISBN 978-1-4333-6976-6

Printed in China. Jinhao. 466494

Table of Contents

Knowing Places

You know where things are in your room and home. Can you get around other places, such as your school?

This is a school in Chicago.

Maps

How do you get from one place to another? A **map** can show you the way.

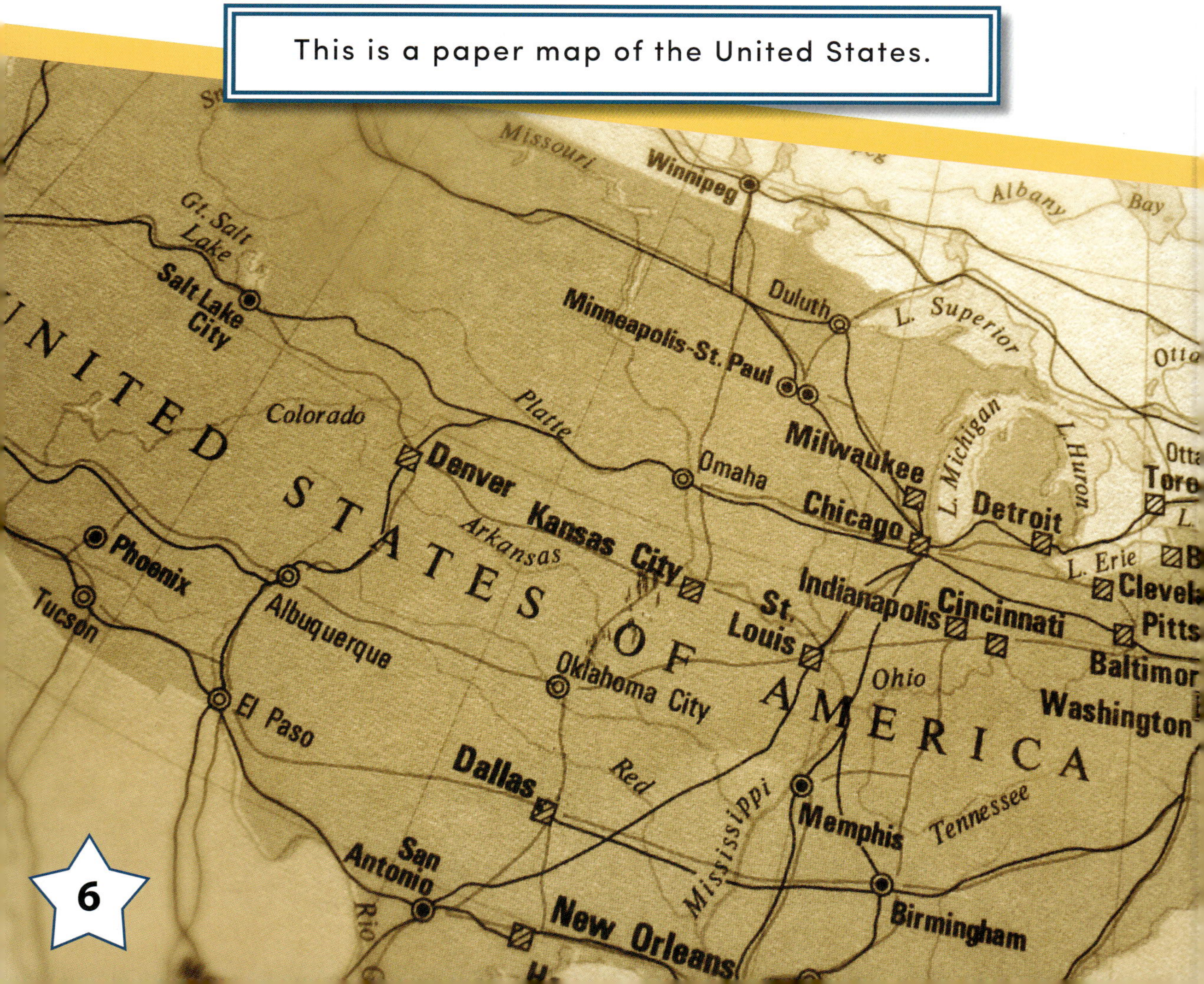

This is a paper map of the United States.

Use a Computer

A **navigation** (nav-i-GEY-shuhn) system can read a map for you. It can tell you where to go.

This is a computer map in a navigation system.

Around School

This map shows places at a school. It uses shapes and lines. It also uses **symbols** (SIM-buhlz).

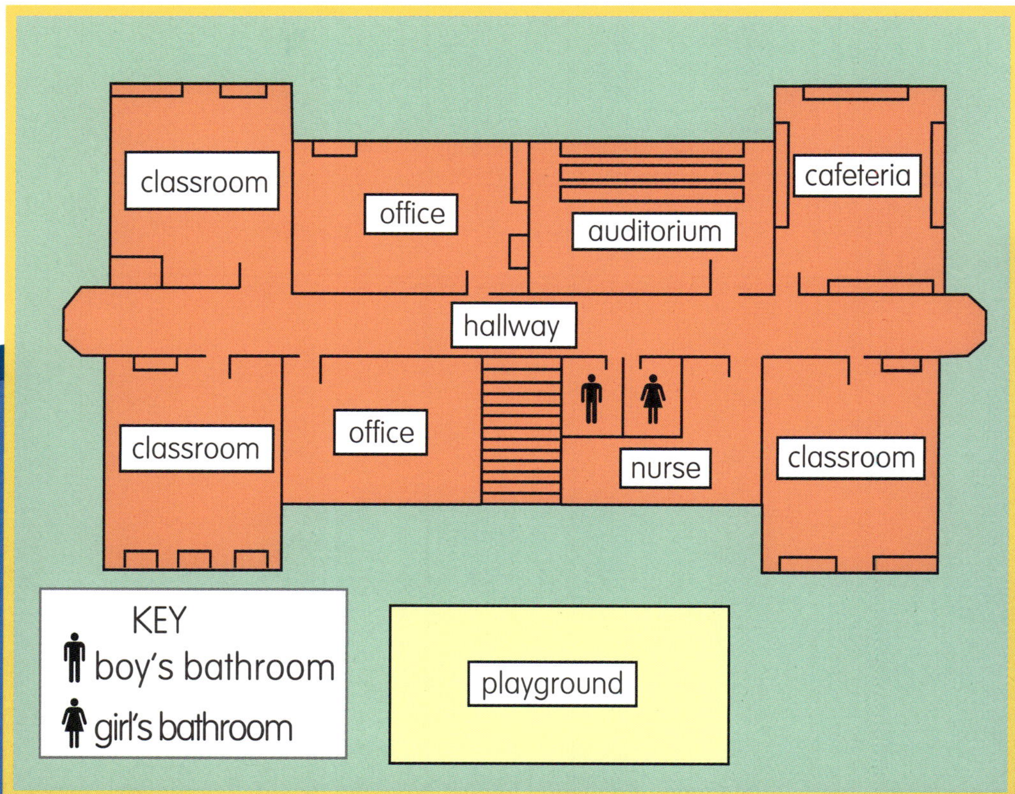

This is a map of a school.

Map Key

Maps use symbols. A map **key** tells what each symbol means.

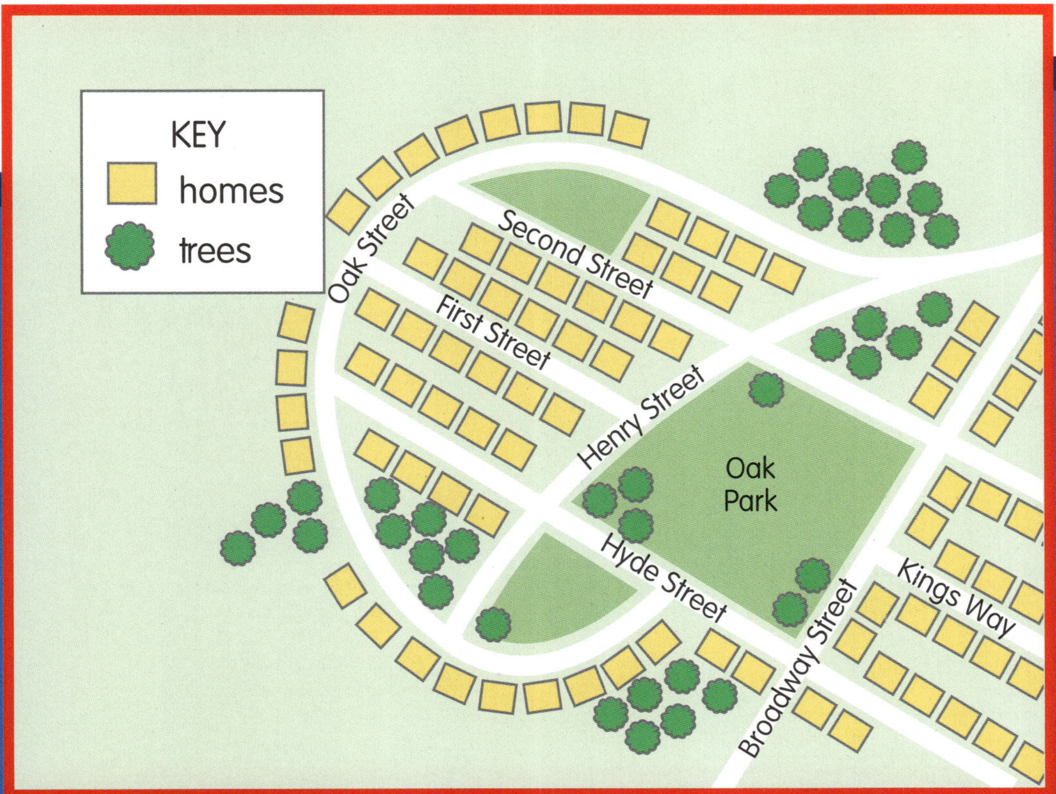

This map has a key at the top.

Around the Classroom

This map shows a classroom. The shapes show where different things are in the room.

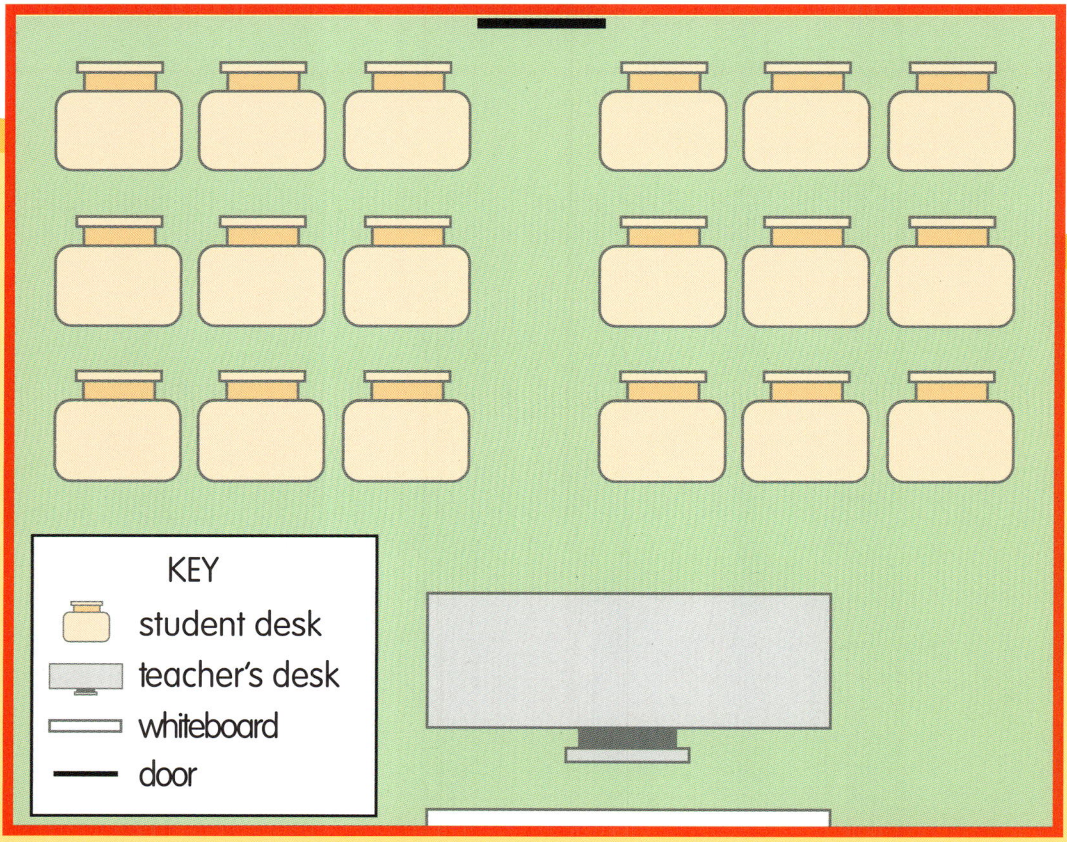

classroom map

This is a classroom today.

This teacher stands in **front** of the board. The students sit **behind** their desks.

This is a classroom in 1942.

Look and See

Look forward to see what is in front of you. Look back to see what is behind you.

A boy looks to see what is behind him.

Each thing has a place in the classroom. The flag hangs **above** the board. The board is **below** the flag.

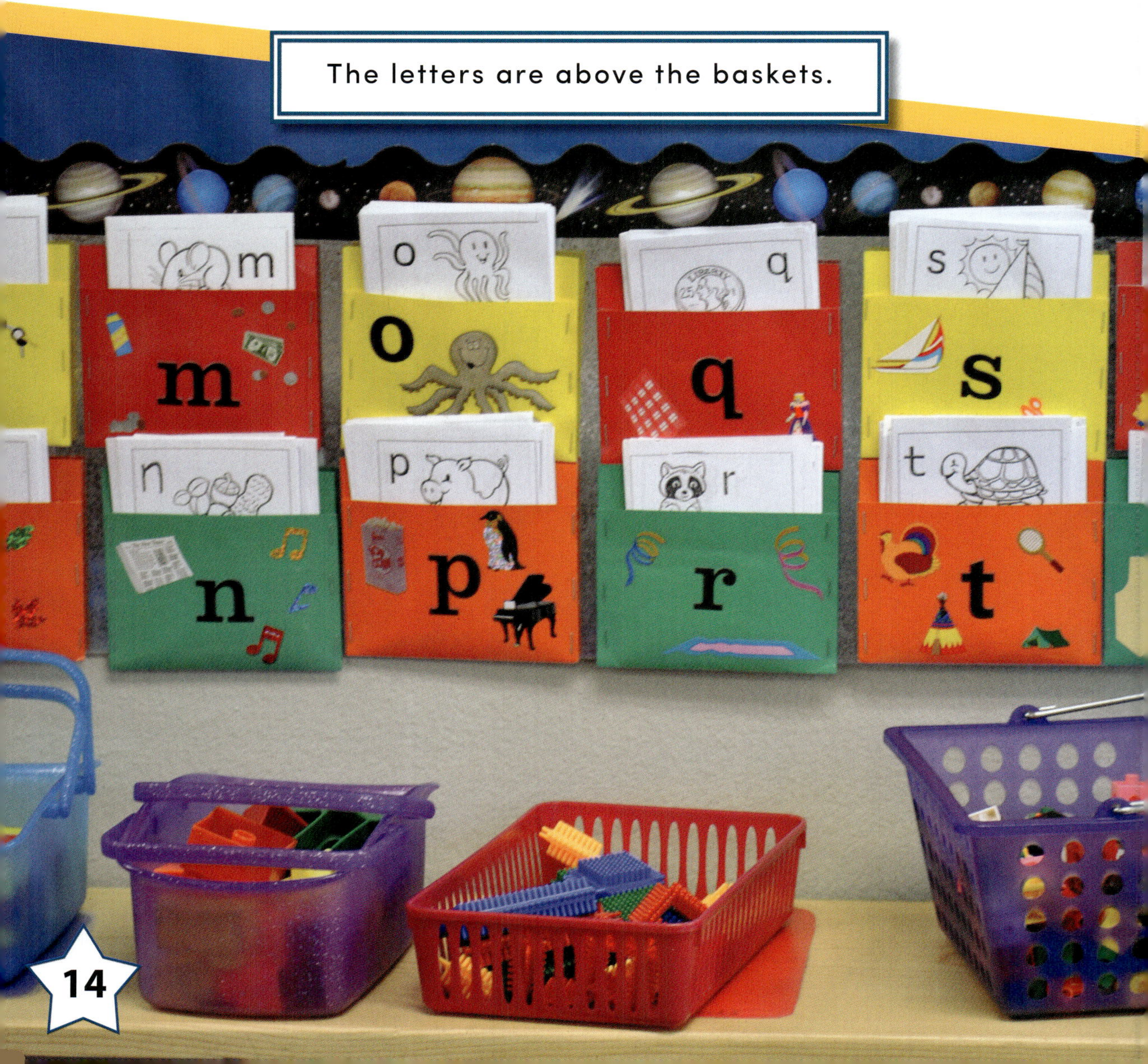

The letters are above the baskets.

The clock is above the flag.

Find Your Way

A **compass rose** shows north, south, east, and west. Use these directions to find your way.

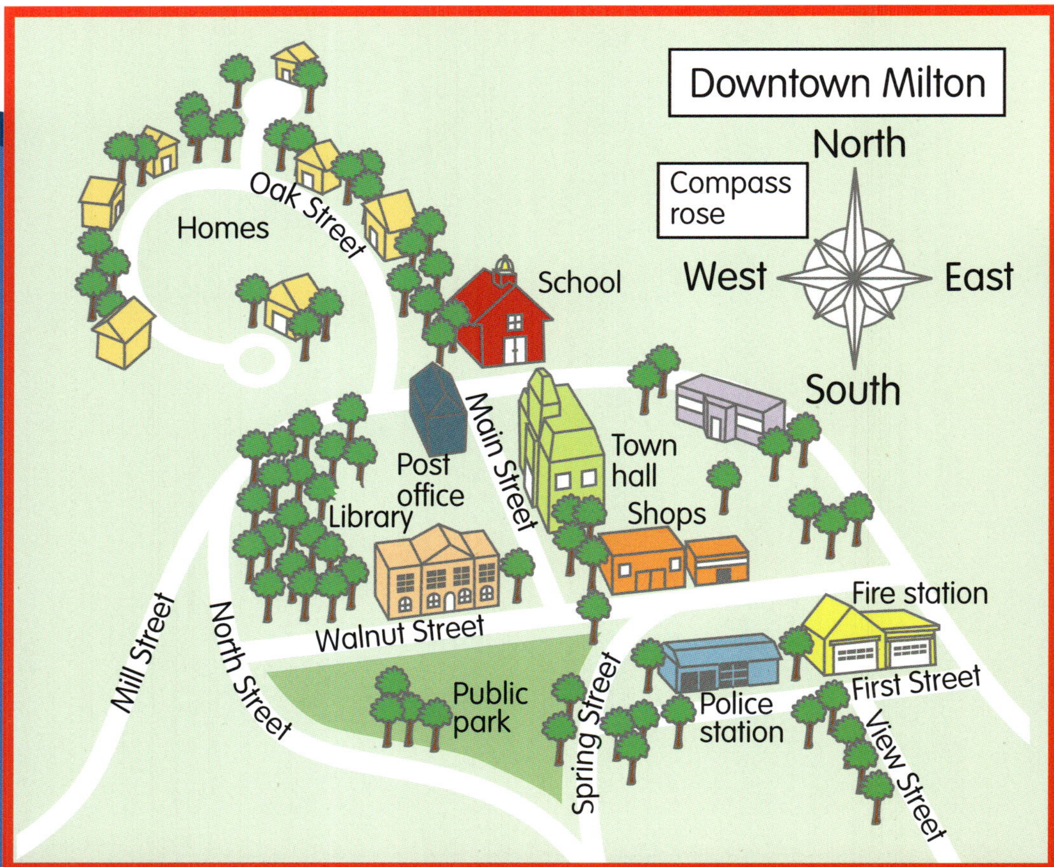

Can you find the compass rose on this map?

Wind Rose

The compass rose was first called a *wind rose*. It showed wind direction.

This is a compass rose.

Useful Tools

Students need to know how to get around school. Maps can help! Directions can help, too! They are useful tools.

The map helps the boy find his way.

This girl uses a map to find special places near her school.

Find It!

Hide an object in your classroom. Write directions or tell a friend how to find it. Use the words from the word bank in your directions.

Word Bank

above
behind
below
east
far
front
left
near
north
right
south
west